Disclaimer:

All the material contained in this workbook is provided for educational and informational purposes only. No responsibility can be taken for any results or outcomes resulting from the use of this material.

While every attempt has been made to provide information that is both accurate and effective, the author does not assume any responsibility for the accuracy or use/misuse of this information.

Printed in the United States of America

Published by:

Lasting Press

615 NW 2nd Ave #915

Canby, OR 97013

For more information about George Mansour or to book him for your next event, speaking engagement, podcast or media interview, please visit GeorgeMansour.com

Part I
Awakening

"What does it mean to be human in the age of technology?"
—Tom Chatfield

Internet Habits Questionnaire

1. How often do you use the Internet?

2. How many hours a day do you spend online?

3. How often do you check your email?

4. How often do you check social media?

5. What are the primary reasons you go online?

6. In what ways do you feel using the Internet has improved your quality of life?

7. In what ways do you feel using the Internet has diminished your quality of life?

8. Do you feel you have a strong understanding of how the Internet works?

9. What steps are you currently taking to keep you and your data safe?

10. Do you feel safe online?

Technology Inventory

List all of the technology you have in your home or workplace that connects to the Internet. Think carefully about all of the devices you use. I recommend walking through your environment and paying close attention to each device.

Device - What is it?	Purpose - What do you use it for?	Usage - How often do you use it?
Example: Fitbit	*I use it to track my physical activity and monitor my health.*	*I use it 24 hours a day. I check it multiple times a day.*

Reflection Questions

How does this awareness make you feel? Do you need all of these devices? How do these devices improve your life? How are they negatively impacting your life?

_____

_____

_____

_____

_____

Commonly Used Websites and Applications

List all of the websites you visit regularly and the applications you use often. Review your history to see what sites you've been using. Go through your phone and note every application you've downloaded.

Website/Application - What is it?	Purpose - What do you use it for?	Usage - How often do you use it?
Example: Facebook	I use it to stay in touch with friends and family and share information about my life.	I check it multiple times a day.

Reflection Questions

How much time are you spending online? Do you feel it is too much? How does this online activity improve your life? How does it negatively impact your life?

_____

_____

_____

_____

_____

Part II
Dangers of Technology

"Technology is a useful servant but a dangerous master."
—Christian Lous Lange

Vulnerability Identification

Identify the greatest threats to you, your family, and your business. What are your vulnerabilities? Pinpoint the specific sites, apps, and devices that are drawing you in, gathering your data, and putting you at risk.

Is your time spent on social media putting you in danger? Answer these questions before you continue to use social media:

1. Are my profiles public or private?

2. What personal information have I shared on my profile?

3. Who am I connected to on social media?

4. Do I answer quizzes found on social media sites?

5. Do I click on links from social media sites?

6. Do I update the password for my account?

7. Do I link my accounts to other sites or applications?

Are your applications putting you in danger? Answer these questions before you continue to use apps on your phone or tablet:

1. Do I read the Terms of Service and Agreements before downloading applications?

2. Do I use password protection to avoid unwanted purchases?

3. Do I link my apps to other online accounts?

Are your emails putting you in danger? Answer these questions before you send or transmit any content data:

1. What content data is being accumulated and stored?

2. When will my content data get cleared?

3. Where can I view my content data?

4. Who has access and permissions to my content data?

5. Who is making decisions based on my analytical content data?

6. Why is my content data being profiled?

7. How is my content data being delivered, used, and shared?

Signs of Internet Dependency

Internet Addiction is a growing epidemic. The questionnaire below provides information on identifying and treating this type of addiction. Answer the following questions honestly. If you answer yes to five or more, you may need support in using the Internet in moderation.

1. Do you feel preoccupied with the Internet (think about your previous online activity or anticipate your next online session)?

2. Do you feel the need to use the Internet with increasing amounts of time in order to achieve satisfaction?

3. Have you repeatedly made unsuccessful efforts to control, cut back, or stop Internet use?

4. Do you feel restless, moody, depressed, or irritable when attempting to cut down or stop Internet use?

5. Do you stay online longer than originally intended?

6. Have you jeopardized or risked the loss of a significant relationship, job, educational, or career opportunity because of the Internet?

7. Have you lied to family members, therapists, or others to conceal the extent of your involvement with the Internet?

8. Do you use the Internet as a way of escaping from problems or of relieving a dysphoric mood (e.g., feelings of helplessness, guilt, anxiety, depression)?

Beard, Keith W. "Internet Addiction: A Review of Current Assessment Techniques and Potential Assessment Questions." *CyberPsychology & Behavior*, vol. 8, no. 1, 2005, pp. 7–14., doi:10.1089/cpb.2005.8.7.

Part III

Developing the Unhackable Mindset

"If you change the way you look at things, the things you look at change."
—Wayne Dyer

Data Protection Implementation

Begin shifting your mindset about how you use technology. Your transformation is not going to happen overnight. Review these cybersecurity suggestions from Chapter 11 - Steps to Securing Your Life and Business. Identify the first five steps you will implement to protect your data.

Cybersecurity Steps that Put Control Back in Your Hands:

1. Avoid Data Sync Programs

2. Protect Your Big Data

3. Carefully Install Programs and Remove Bloatware Software

4. Backup and Archive Policies

5. Limit the Use of Free Platform Environments like Google and YouTube

6. Short-Term Gains or Rewards to Stay Away From

7. Think Twice Before Using New Technology

8. Read Over Agreements

9. Focus on Verification and Validation

10. Stop Smartphone and Device Tracking

11. Improve Password Practices

12. Avoid Public Wi-Fi

13. Security Assessment & Education and Training Awareness

14. Use a Better Email Address

15. Use a Flip Phone, Light Phone, or a Custom Rom on weekends

16. Update Your Technology

17. Delete Activity Controls

18. Don't Let Any Products or Services Give You a False Sense of Security

19. Don't Fall for Spam Phone Calls and Text Message Scam

20. Be Careful When Using Apps and Trusted Devices

21. Use Facebook and Other Popular Social Media Sites Carefully

It is not possible to implement all 21 steps at once; however, over time you can make all of these changes happen and improve the way you interact with technology. As you begin this shift, commit to focusing on five of the above steps starting today.

Step 1: What step will you take? What support do you need to take this step? Why have you chosen this step?

_____

_____

_____

Step 2: What step will you take? What support do you need to take this step? Why have you chosen this step?

_____

_____

_____

Step 3: What step will you take? What support do you need to take this step? Why have you chosen this step?

_____

_____

_____

Step 4: What step will you take? What support do you need to take this step? Why have you chosen this step?

_____

_____

_____

Step 5: What step will you take? What support do you need to take this step? Why have you chosen this step?

_____

_____

_____

Bonus Security Resources

→ Should I Remove It? (www.shouldiremoveit.com) "is a very simple but extremely powerful Windows application that helps users, both technical and non-technical, decide what programs they should remove from their PC. This typically includes finding and removing all sorts of crapware and bloatware such as adware, spyware, toolbars, bundled unwanted applications as well as many forms of malware."

→ CCleaner (www.ccleaner.com) Basically like hiring a Merry Maid for your computer! As we move through our day of computer use, we really have no idea how much junk in the way of cookies, browsing data, and trackers are taking up space on our computers and slowing them down. That free, fresh feeling you get after a deep house cleaning will come after using this site – you'll feel revitalized using a faster, smoother running computer and safer, knowing that when you go online, nothing is tracking your online movement.

→ Revo Uninstaller Pro (www.revouninstaller.com) "is an innovative uninstaller program that helps you to uninstall unnecessary software and remove easily unwanted programs installed on your personal computer."

→ DuckDuckGo (www.duckduckgo.com) is a great search engine that doesn't collect or share any of your personal data. They don't track you or store your search history. "Too many people believe that you simply can't expect privacy on the Internet. We disagree and have made it our mission to set a new standard of trust online."

→ Startpage (www.startpage.com) "gives you web search results from Google in complete privacy. When you search with Startpage, we remove all identifying information from your query and submit it anonymously to Google ourselves. We get the results and return them to you in total privacy."

→ HMA (www.hidemyass.com) "HMA VPN strips away everything that makes you unique and identifiable online, hiding your IP address and encrypting your data to keep your browsing history private."

→ Nomorobo (www.nomorobo.com) robocall protection is now available for both iOS and Android devices.

→ RoboKiller (www.robokiller.com) automatically blocks telemarketers and robocalls from ringing, even if they are spoofing or changing their numbers.

→ Hiya (www.hiya.com) provides spam detection and call blocking.

→ Proton Mail (https://protonmail.com) is "an easy to use secure email service with built-in end-to-end encryption and state of the art security features."

➔ Here are some sites for disposable emails: www.10minutemail.com, www.throwawaymail.com, www.mailinator.com, www.tempmailaddress.com

➔ Tor browser is portable and runs off a USB flash drive and has a pre-configured web browser to protect your anonymity. "The Tor software protects you by bouncing your communications around a distributed network of relays run by volunteers all around the world: it prevents somebody watching your Internet connection from learning what sites you visit, it prevents the sites you visit from learning your physical location, and it lets you access sites which are blocked." (www.torproject.org/projects/torbrowser.html.en)

➔ Ninite (www.ninite.com) is a package management system offering that lets users automatically install popular applications for their operating system. The easiest, fastest way to update or install software. Ninite downloads and installs programs automatically in the background. It will package, install, and update all your programs at once with no toolbars or clicking next, so just choose your applications.

➔ OpenVPN (www.openvpn.net) is a free open-source software that configures your Virtual Private Network (VPN) so that you can create secure point-to-point or site-to-site configurations as you access and work remotely and/or if you need different sites securely connected.

➔ SHADE Sandbox (www.shadesandbox.com) is an alternative for antivirus and a tool for virtualization.

➔ Sandboxie (www.sandboxie.com) is isolation technology to separate programs from your underlying operating system, preventing unwanted changes from happening to your personal data, programs, and applications that rest safely on your hard drive.

➔ Windows Sandbox (www.thewindowsclub.com/windows-sandbox) is a temporary desktop environment where you can run untrusted software without the fear of a lasting impact to your PC. Any software installed in Windows Sandbox stays only in the sandbox and cannot affect your host. Once Windows Sandbox is closed, all the software with all its files are permanently deleted.

➔ Speccy (www.ccleaner.com/speccy) "Fast, lightweight, advanced system information tool for your PC. Need to find out what's inside your computer? Speccy has all the information you need!"

➔ HWiNFO (www.hwinfo.com) "Professional System Information and Diagnostics. Comprehensive hardware analysis, monitoring, and reporting for Windows and DOS."

➔ Find out if any of your accounts have been compromised in any of the data breaches at www.haveibeenpwned.com

➜ HTTPS Everywhere is a security plugin that encrypts all communication across the web to help secure your browsing experience. Find it at www.eff.org/https-everywhere

➜ Anonabox adds another layer of protection to your online connection by adding more privacy and anonymity by deterring big data collection and analysis, deters hackers with large barriers, anonymous wireless VPNs, keeps your location secret by changing your location, online browsing privacy, encrypting your data (VPN), stops remarketing ads that are location specific that retarget you, connects multiple devices and you can access location-specific content. Find it at www.anonabox.com

➜ CheckShortURL (www.checkshorturl.com) helps you verify the integrity of your shortened links to detect malicious activities, such as phishing attacks, malware, viruses, or inappropriate content for work that is not safe or spam related. You will be able to see the original URL from the shortened link before you tap on anything. It allows you to see the full link before you decide to move forward as it shows you where it's pointing to. These are two similar sites: www.trueurl.net, www.urlrevealer.com

➜ Check a website or URL for phishing, malware, viruses, and reputational issues. Here are a few of the sites to check for safe browsing: www.urlvoid.com, www.scanurl.net, https://safeweb.norton.com, https://sitecheck.sucuri.net

➜ PhishTank is one of many ways to check to see and verify if a site link is a phishing scam, which is usually a fraudulent attack made through email to steal your personal information. PhishTank will alert you and help you spot phishing emails and sites. Find it at www.phishtank.com

➜ JavaScript can be abused and isn't perfect. This makes it possible to snoop on your internet activity and violates your privacy. It also makes it easier to track your behavior. Most of the big-name sites would be nowhere near as useful if JavaScript did not exist. If your aim is privacy and anonymity, then you should make sure you block JavaScript and disable it completely. If it becomes problematic, you can then turn it on for the site you need it on and then shut it off once completed. You can install the NoScript extension for Firefox and ScriptSafe for Chrome users. Both choices will allow you to decide which sites it should load JavaScript for. (www.noscript.net, www.github.com/andryou/scriptsafe)

➜ IPVOID is one of many IP Address tools online that will help you discover details about a specific IP Address. (www.ipvoid.com)

➜ DISCONNECT will keep your data safe, secure your identity without selling or tracking your data, break loose from all unwanted tracking online, and enjoy a faster, safer internet all the time as you browse online. (www.disconnect.me)

➜ DNS leak will make sure that once you're connected to a VPN, you are only using the DNS servers provided by that VPN service provider. Instead of your local ISP network provider where they can log and track your activity. (www.dnsleaktest.com)

→ Panopticlick checks to see how your browser and add-ons protect you against online tracking tactics. It also checks to see if the system is identifiable or uniquely configured with a fingerprint, independent of whatever privacy software you're using. This is a great tool to protect your anonymity and test how well your privacy protection and actions are working. (www.panopticlick.eff.org)

→ OnionShare shares files of any size securely and anonymously. It's a free open source tool. No third party or separate server file-sharing service is needed. This is a great alternative instead of using any cloud storage services. (www.onionshare.org)

→ IronKey delivers one of the highest standards of mobile storage protection available. You can lock down sensitive data with an on-board, hardware encryption engine that puts a military-grade wall between unauthorized users and the drive's contents. It has a variety of flash drives and external hard drives. (www.ironkey.com)

→ Tails is a live operating system that aims to preserve your privacy and help you become anonymous online. You can start it from any computer using a USB media device or a DVD. It will allow you to use the Internet in an anonymous manner and overcome any filters by forcing all your online connections to go through the incognito Tor network. It will leave no trace on your device unless you ask it to. It incorporates state-of-the-art encryption tools for all your files, messaging, and emails. (tails.boum.org)

→ Qubes OS is a free, security by compartmentalization operating system. It divides everything into specific categories called Qubes. It all works in its own secured isolated container performing its intended function and purpose. It's a good defense to guard against your own unconscious psychological habits, behaviors, and patterns. This way if one Qube gets compromised, it won't affect the others or try to comprise your entire operating system. You're able to keep different things you do secure and separated. You can create one Qube for all your public browsing and another as a disposable Qube for all your private browsing that can be created from a template quickly and will disappear when closed. (www.qubes-os.org)

→ Privacy Badger blocks all invisible trackers as you surf the web and is also able to enable a do not track option if desired. It's a browser add-on that stops advertisers and third-party trackers from tracking your online activity, like knowing where you go and what page you're looking at. (www.eff.org/privacybadger)

→ Reputation Defender will help you clean your personal data online and help you become more aware by controlling the search results for your name and removing your personal information from the internet.(www.reputationdefender.com)

→ Limit the information that is out there by removing your presence from online sites. You can use a service site like Delete Me (www.deleteme.com) to assist you.

Part IV

A Paradigm Shift

"A simple paradigm shift is all it takes to change your life forever."
—Jeff Spire

Now that you have an Unhackable Mindset, who are the first five people you are going to share it with?

1. _____

2. _____

3. _____

4. _____

5. _____

How are you going to help your business, family, and community make more mindful choices around cybersecurity?

_____

_____

_____

What are the next steps you are going to take to become your own security shield?

_____

_____

_____

In what ways can you support others with their Unhackable mindset transformations?

_____

_____

_____

What additional support and resources do you need in order to further develop your Unhackable mindset?

_____

_____

_____

Additional Resources

The following are resources to help you as you continue to develop an Unhackable mindset. Below you'll find some of my top resources for taking your cybersecurity to new heights.

Websites

Center for Humane Technology (formerly called Time Well Spent) - "a nonprofit addressing the harmful extractive attention economy by inspiring a new race to build Humane Technology that aligns with humanity." www.humanetech.com

Common Sense Media - "the nation's leading nonprofit organization dedicated to improving the lives of kids and families by providing the trustworthy information, education, and independent voice they need to thrive in the 21st century." www.commonsensemedia.org

Electronic Frontier Foundation (EFF) - "the leading nonprofit defending digital privacy, free speech, and innovation." www.eff.org

Electronic Privacy Information Center (EPIC) - "EPIC was established in 1994 to focus public attention on emerging privacy and civil liberties issues and to protect privacy, freedom of expression, and democratic values in the information age." www.epic.org

Films

Screenagers - An award-winning film that probes into the vulnerable corners of family life and depicts messy struggles over social media, video games, and academics. The film offers solutions on how we can help our kids navigate the digital world. www.screenagersmovie.com

Captivated: Finding Freedom in a Media Captive Culture - It addresses the potential physical, moral, spiritual, mental, and emotional impact of today's media and technology when consumed or used without discretion. www.captivatedthemovie.com

About The Author

George Mansour has a unique and simplified approach to cybersecurity. As a trusted IT specialist for more than twenty years, he helps clients overcome the debilitating effects of cybersecurity issues. His distinctive methods have allowed him to emerge as an industry leader for individuals and businesses seeking to find a more secure system.

George understands the imminent threat facing anyone connected online. His goal is to empower end-users with the tools to secure their digital assets using proper protocols. He seeks to enhance the validity of his client's lives, by teaching them the basics of online security. The systems, products, and services he uses are designed to prevent a breach before it happens. Preventative methods are always preferred. It is imperative that users become proactive, because cognitive distortion has taken over our senses. We need a compass to navigate the Internet in real-time, and we need to develop a license agreement with ourselves that isolates our unique, sensitive data.

George received a Bachelor of Science in Business Administration (BSBA) degree with a major in Computer Information Systems (CIS) from Suffolk University, along with a Microsoft Certified Professional (MCP) certification. Besides his academic accomplishments, he is the founder and owner of CEHIT, INC., an acronym for Computer Engineering Hardware Information Technology —an Information Technology company that has helped manage complex IT environments for over 15 years. CEHIT, INC. has been helping thousands of consumers and businesses (small, medium, and large) across the globe. George Mansour focuses on numerous vertical markets and organizations in the healthcare, law, insurance, manufacturing industries, etc. He has affiliations with partners and value-added reseller programs. Cybersecurity is a shared responsibility.

George's message is that technological interaction affects everyone. We are interwoven over the Internet as one, but individually unique. The Unhackable mindset creates the conceptual model to enact a multi-layered digital union between technology and human interaction. The insecurities and impurities that threaten us online with every new connection must be controlled through a collaborative approach that will suppress threats to our sensitive data. He says it's time to reclaim our digital lives and our digital freedom.

Hearing George's message and getting his unmatched advice and step-by-step approach to cybersecurity is something everyone should immediately add to their to-do list! Don't miss out on your opportunity to work with George and be a part of this tech paradigm shift. Visit www.GeorgeMansour.com to learn more about becoming Unhackable.

www.ingramcontent.com/pod-product-compliance
Lightning Source LLC
Chambersburg PA
CBHW080634030426
42336CB00018B/3190